Interconnectedness of Life

By
Moses Ekebuisi

Grosvenor House
Publishing Limited

All rights reserved
Copyright © Moses Ekebuisi, 2025

The right of Moses Ekebuisi to be identified as the author of this
work has been asserted in accordance with Section 78
of the Copyright, Designs and Patents Act 1988

The book cover is copyright to Moses Ekebuisi

This book is published by
Grosvenor House Publishing Ltd
Link House
140 The Broadway, Tolworth, Surrey, KT6 7HT.
www.grosvenorhousepublishing.co.uk

This book is sold subject to the conditions that it shall not, by way of
trade or otherwise, be lent, resold, hired out or otherwise circulated
without the author's or publisher's prior consent in any form of
binding or cover other than that in which it is published and
without a similar condition including this condition being
imposed on the subsequent purchaser.

A CIP record for this book
is available from the British Library

ISBN 978-1-80381-535-0
eBook ISBN 978-1-80381-536-7

Dedication

This book is dedicated to all those who play roles in facilitating nature conservation, human freedom and welfare, and peace of the world. Their efforts may seem unnoticed, unappreciated and unrewarded; however the message from this book has illuminated the significance of their endeavour and revealed the subsequent hidden recompense. Their work is now visible as contributing to the protection, conservation, and care of the building blocks that make up and sustain the sacred universe and its order – a divine duty that earns rich and everlasting spiritual and eternal reward.

Acknowledgements

My acknowledgements go to the following people for their support, help and encouragement which made it possible for this book to materialise:

Professor J. S. Bale – Head of Science Research; Chris Wright – Chief Scientist; W. A. Thomson – Departmental Superintendent.

I also acknowledge Dr. Jim Newmark for general proof reading and citation. Further, I would like to acknowledge Dr. Chamberlain Agbada for data inspection and presentation, and proof reading of the annex (the scientific aspect of the book).

Table of Contents

Introduction	ix
The Author's Biography	xv
Chapter 1 Interconnection & Interdependence of Soil Organisms	1
Chapter 2 Interconnection & Interdependence of Soil organisms and other Lifeforms	14
Chapter 3 Interconnectedness of Life, a Key to a Better Life and Lasting Success	21
Chapter 4 The Tragedy of Failure of Man's Roles in the Interconnectedness of Life and Its Models	25
Chapter 5 Lessons Learnt and Insight Gained	42
Annex (Report of the Research Work and the Results)	51

Introduction

Interconnectedness of life is a concept that considers that the life of all living things, including man, is interconnected and interdependent, and that no living thing can exist in isolation, independent of the others.

For thousands of years this idea has been held by different groups of people. It was expressed in Gaia theory. It is part of Buddhists' belief. Expression of this thought is in the Christian holy book, the Bible. Most scholars agree that all living things are interconnected and dependent on each other in such a way that a change in one lifeform cannot be fully understood without considering the simultaneous changes in the others.

Ecologists see this interconnectedness of life as an expression of the wisdom of nature. Mystics and religious groups perceive it as an expression of the wisdom of God the Creator. Whichever way one views it, one commonly held fact about it is that it is a system that exists as a reflection of the immense wisdom of any entity from which it originated.

It was through a mixture of mystical experience and scientific findings that I, the author proved the existence of the interconnectedness of life. The details of these are in chapter 1.

The whole thing began when I was involved in research in an area of agricultural science in which I played a major role. This research took place at the Headley Hall Farm, under the auspices of Department of Pure and Applied Biology, Leeds University in the United Kingdom (UK). It was started in 1987 and completed in 1989.

The aim of the research was to find out how agricultural pesticides affect the pests of field crops and their natural enemies, and how the chemicals also affect some grassland soil organisms known as microfauna.

As the work progressed, we started noticing an unanticipated and mysterious occurrence. The occurrence was consistent and very noticeable in permanent grassland. On some occasions, in the plots where the pesticides were used, the subsequent grass yield tended to be higher than where the chemicals were not applied. We checked if the higher grass yield was because the chemicals saved the yield by killing pests in the plots with the pesticides. We found out that this was not the reason. There was no significant difference in pest population between the plots with the pesticides and the ones without them. So, this strange occurrence remained a mystery.

It was during this period that a mystical event occurred and gave the key that unlocked the mystery. I was the one who had the experience.

The message I received is that the reason behind the finding is because:

> *'The animals supply the needs of man;*
> *the plants supply the needs of the animals;*

for all the soil organisms, bits and pieces assembled together, mimic the ecosystem in order to supply the needs of the plants'.

The ecosystem in the soil is a system where groups of the soil organisms are connected in an interdependent manner. In the system, the population of each group of the organisms stands in delicate balance with one another.

The message I received led to my identifying how the soil organisms and plants are interrelated. I found out how the organisation and functions of the soil organisms in the ecosystem result in provision of nutrients for plants in an appropriate and regulatory way, just as the plants provide food for animals.

I identified that the pesticides affected the balance of the populations of the groups of the soil organisms in the ecosystem in such a way that the result was the increase in the grass yield. (The details of this are in chapter 1).

In chapter 2 is more scientific evidence showing that the soil organisms and the plants are not only interrelated but also are connected with other lifeforms in the soil and above the soil. In this chapter is scientific proof of how all organisms, including the soil organisms, are interdependent in such a way that one cannot exist without the existence of the other. For example, a process known as photosynthesis illustrates how plants and animals are interdependent.

In this chapter is also evidence of interdependence of land and sea lifeforms, despite the fact that land is separated from the sea.

Chapter 3 is an assemblage of ideas of scholars and mystics which agree with each other that an insight into the principles of the system of interconnectedness of life is a key to great wisdom and its practical application.

Chapter 4 is an analysis of the attributes with which we, humankind evolved as our unique characteristics, the characteristics of the caretakers and managers of the system of the interconnectedness of life. These attributes are identified as directors of our life, regulators of our attitude and behaviour so that we live, function and play our roles in accordance with the principles of interconnectedness of life, the wise principles upon which our well-being and success depend.

Some of these attributes are visible as humankind's mystical perception, love, conscience, instinct to socialise and accept diversity; our culture and the cultural practice. A common suggestion is that the major cause of humankind's social, health and spiritual problems today can be traced to a loss of one or more of these attributes.

Also in this chapter is a summary of my own work and research which show that music is very important in a culture for enhancing social connection and improving well-being.

In chapter 5, the lessons learnt and insight gained from the knowledge of interconnectedness of life are pulled together. It is shown here that religion, a form of culture with which humankind evolved, is a model of interconnectedness of life for achievement of our spiritual needs and for gaining eternal life.

This chapter brings together the teachings of the Buddhists; the teachings of St Paul, one of the most important figures in the apostolic age; and the message of Jesus Christ, the central figure in Christianity. The teachings and the message all relay the same thing – that to achieve spiritual peace and success and earn eternal life, one has to live a life that is in tune with the principles of interconnectedness of life. Among the principles are one's expression of love and acceptance and treatment of people with their unique characteristics as good neighbours – as much as those who are related to us.

Finally in this chapter, the book ends with this message about the knowledge of the system of interconnectedness of life:

> *'The understanding of the system has made it clear that to be able to beat our swords into ploughshares and our spears into pruning hooks, and achieve lasting world peace and order, we need to model the world on the structure and principles of interconnectedness of life; otherwise, we are engaging in a perpetual task with limited success.'*

This book is meant to be readable by everyone, including those with no science background. I have therefore tried to avoid putting too much scientific information in it. However, I have reserved the Annex for basic scientific information that is applicable to the research. So, the information in the Annex just repeats, in a simple scientific way, what is already said in some chapters of this book. So, it is not necessary to read it if one has already read all the chapters (chapter 1 to 5).

In the Annex are the overview of the research procedure, the research data, and the graphic representations of the data.

The Author's Biography

Moses Ekebuisi was born on February 1960 in Ngodo Isuochi, Umunneochi Local Government, Abia State, Nigeria.

Moses, a Biomedical Sciences graduate of Bradford University in the United Kingdom (UK) has been working in the field of science and arts for a long time. In 1986 he worked in mosquito research for Leeds City Council in the UK.

From 1987 to 1989, he worked with scientists at Leeds University in the UK in research to investigate the effect of seed bed preparation, soil structure and release time on the toxicity of a range of grassland pesticides to the carabid beetle. The result of the research was published in the Journal of Applied Entomology[1] and Moses is one of the co-authors of the paper.

Further research in the University investigated the effect of pesticides on crop pests in grassland and on some

---

[1] J. S. Bale, M. Ekebuisi and C. Wright. "Effect of seed bed preparation, soil structure and release time on the toxicity of a range of grassland pesticides to the carabid beetle Pterostichusmelanarius (Ill.)(Col., Carabidae) using a microplot technique." *Journal of Applied Entomology* 113.1–5 (1992): 175–182.

organisms (microfauna) in the grassland soil. It was both these scientific findings and Moses' mystical experience in this particular work that inspired him to write this book.

Moses is also the author of the novel, 'Man at Nature's Pinnacle', a novel about the importance and significance of man's interrelationships with animals and plants, as demonstrated in ancient Europe. It was the experience he gained from the research that inspired him to write the novel. The novel and this book are available in all major book retailers.

Other research followed involving Moses and scientists in the Clinical Oncology Unit, University of Bradford. The aim of the research was to assess the chemosensitivity of cancer cells to indoloquionone compound E09 which was selected for phase 1 evaluation by the European Organisation for Research and Treatment of Cancer (EORTC). The result of the work was published in the Journal, Annals of Oncology, and Moses is one of the co-authors.[2]

Later, Moses set up a music company, the main purpose of which was to research and identify if multicultural music helped in promoting community cohesion and development; reducing social isolation; and developing creative, social, educational and life skills.

---

[2] Phillips, R. M., M. C. Bibby, B. P. Cronin, and M. Ekebuisi. "Comparative in vitro chemosensitivity of DLD-1 human colon adenocarcinoma monolayers and spheroids to the novel indoloquinone EO9." *Ann Oncol* 3 (1992): 106.

The research was successful. Following this, Moses received a financial award from the Millennium Commission to work in communities and schools. Through the work, young people who were excluded from schools gained improvements and returned to mainstream schools to continue their education; concentration and learning skills of special educational needs pupils were improved, leading to their performing better in their school subjects; social isolation of vulnerable people was reduced; and social barriers between different groups were broken, leading to the social integration of the people, improvement of their well-being and their participation in celebration of diversity.

Following the above success, Moses was awarded a Millennium Fellowship by the Millennium Commission. Some of the successes of the work and the award were reported in the Yorkshire Post. The report is still on the internet and can be accessed by typing "*Moses Ekebuisi, Yorkshire Post 2001*", in Google.[3] (or visit his website: www.mosekemusic.co.uk and click on the PRESS link to read the press report).

The year 2000 was the year of the Millennium celebration. It was the year the Millennium Flame toured various cities and events took place throughout the United Kingdom. The flame was used by Her Majesty the Queen to light the National Beacon on December 31st 1999. As part of the tour, the flame was carried from the Tower of London and used to light the first British Legion

---

[3] (http://www.yorkshirepost.co.uk/news/main-topics/local-stories/giftedmoses-bangs-the-drum-for-world-music-1-2418996, accessed November 2015)

Torch of Remembrance which was carried through the streets of London as part of Lord Mayor's show that year. Following this, the torch was moved to Queen Victoria Memorial where it was officially received by His Royal Highness, The Prince of Wales, before the torch was carried to the Festival of Remembrance.

Then, as the flame was about to arrive in Bradford, Moses Ekebuisi was nominated to represent the Bradford community and the Millennium Commission as well as the Millennium Commission award winners. He and the Lord Mayor of Bradford, Cllr. Stanley King, had to welcome the Millennium Flame from Bruno Peek OBE (the founder and Chairman of the Beacon Millennium).

Following the success of his work in the music field, Moses gained a new post as Music Composer Resident at Bradford University.

Since his involvement in the music field, different groups have been accessing his music for promoting diversity; social cohesion; and cultural, ethnic and faith integration. Among the groups were England Football Association which accessed the music for promoting their campaign 'England Against Racism'; The Bradford Inner Ring Group which accessed the music for bringing different faiths together and celebrating achievements.

Based on his research on the therapeutic benefits of musical rhythms, Moses now runs African drumming classes and relaxation sessions for improving well-being. This is in addition to his world music performance.

Chapter 1

Interconnection & Interdependence of Soil Organisms

It is a popular concept that lifeforms on the Planet evolved as a complex network or system where they are interconnected and interdependent, and so it is impossible for one to fully understand changes in one lifeform without considering the simultaneous change in the other.

Tracing back to the distant past, a similar concept was expressed in Gaia theory. Gaia was the primeval Greek mother goddess. The theory proposes that organisms interact with their inorganic surroundings on Earth to form a synergistic and self-regulating, complex system that helps to maintain and perpetuate the conditions for life on the planet.

It was through a research project in which I was chiefly involved that the validity of the concept was proved. In the research I obtained an evidence of interconnection and interdependence of organisms in the grassland soil ecosystem, and I traced how they are linked to other lifeforms.

First of all, let us have a look at what an ecosystem is.

Ecologists define an ecosystem as a biological community of interacting organisms and their physical environment. It is a close-knit unit of which the constituents include plants and animals (including man) together with non-living things.

In the soil ecosystem are various groups of organisms. Each group of the lifeforms has its characteristic population. Some groups are abundant, others less so and some are comparatively rare. Although various factors, including environmental forces, can affect any of the groups' population, making it fall below or rise above its characteristic value, the population normally returns to the characteristic value.

The population of each of the groups of the soil organisms stands in delicate balance with one another. It is through self-regulatory means that the population balance is maintained. The organisms' natural enemies within the system play a part in this. They include predators, parasites, pathogens and herbivores.

If the population of any of the groups starts to rise above the usual level, the natural enemies intensify their regulatory activities; for example, the predators intensify their predatory activities and through this they bring the population down to the normal level. On the other hand, if the population starts to fall below the normal level, the natural enemies relax their action leading to a return of the population to the usual value.

INTERCONNECTEDNESS OF LIFE

In the ecosystem, the groups of the soil organisms are linked in such a way that there are essential flows of energy through a biological food chain among them.

In a simplistic explanation of the flows of energy through a biological food chain, plants such as grass start as the producer of the food. They absorb energy from the sun and with this they are able to make food which is the source of energy for all life. The plants are then eaten by herbivores (the animals that eat plants). Through eating the plants, the herbivores gain the energy. In turn, carnivores or predators (the animals that eat other animals) eat the herbivores and through this they gain the energy. Some other organisms known as decomposers eat the litter that is shed by the plants. They also eat material from dead animals. Through this feeding habit the decomposers gain the energy and in return release mineral nutrients from the organic matter for the plants to absorb, live, grow and continue to make the food. It is by these means that the energy in the food which the plants produce goes round the system.

Interference in the balance of the populations of the groups of the organisms in the soil ecosystem can initiate a chain of changes in the system. An example of this is a chain of changes of which the end result is an alteration of the rate at which the decomposers make the mineral nutrients available to the plants. This was identified in the research in which I was involved. Through the research, I also became aware that the organisms in the soil ecosystem are not just interconnected, but also are part of the myriad of lifeforms in the interconnectedness of all life.

The research was a project in the Department of Pure and Applied Biology in the University of Leeds. The purpose was to identify how agricultural pesticides affect the pests of the field crops and their natural enemies, and how the chemicals also affect the grassland soil organisms known as microfauna.

By means of the research we found out that the pesticides were having lethal effect on many of the grassland soil organisms (microfauna). We also found out that on many occasions in the plots where the pesticides were used, grass yield tended to be higher than in the plots where they were not applied. One could suspect that the high grass yield in the plots with the pesticides could be because the chemicals killed the grass pests in the plots and so saved the grass from damage by the pests. But this was not the reason. We identified that there was no significant difference in pest population between the plots with the pesticides and the ones without them.

It was this strange occurrence that baffled everyone in the research. To find a cause of this, there was an alternative thought of setting up a new project for investigating whether the pesticides were doing something to the biochemistry of the grass.

It was during that period that I received information in an unusual way. This happened one night in my flat, Flat 1 Headley Hall Farm. It was around midnight when I found myself in a deep trance that I had never experienced before. In that state, I heard the head of the project, J. S. Bale calling me in a loud, pressing and excited voice. On my answering the call, he raised up a sheet of white paper which was shining like a dazzling sun. He told me that it was on the paper that the reason behind the higher grass yield in the plots with the pesticides was written. From where I was, I could not see anything written on the paper except the dazzling white light that was coming from it. So, I attempted to move closer to him, but he told me in a loud and powerful voice that I should not come close or attempt to touch the paper because it was too powerful for me. He said that instead I should stay where I was and listen to a voice which was to read out what was written on the paper. So, I listened and I suddenly saw the Leeds University farm secretary. In a clear, simple poetic voice, she read out this:

> "The animals supply the needs of man;
> the plants supply the needs of the animals;
> for all the soil organisms, bits and pieces assembled together,
> mimic the ecosystem in order to supply the needs of the plants."

After this, I woke up. It was still midnight. I wrote the information down and then went back to sleep.

In the research, I was chiefly involved in sampling the soil in the plots, extracting and counting the soil organisms (microfauna). So, from the following morning after the dream, I looked deeper into the data of the soil organisms I sampled and recorded. Of course the pesticides usually reduced the populations of different groups of the soil organisms and in many cases had little or no effect on one particular group known as oribatid mites. We were aware of this. But one thing unnoticed was that in many cases, the pesticides seemed to induce a steady growth of the population of the mites by reducing the populations of the other groups of the soil organisms. I noticed that the more the populations of the other groups were reduced, the more the oribatid mites' population rose, even above the usual level in the plots. I found out that this occurrence is consistent in permanent grassland (established grassland soil ecosystem).

I also noticed that among the groups of the soil organisms on which the pesticides mostly exerted their lethal effects and remarkably reduced their numbers, was a group known as mesostigmatic mites. These mites are well known predators in the soil. I noticed their effectiveness in predation. During one of those times when I was extracting the soil organisms from the soil samples, I came across an occasion where the predatory mites behaved like lions in the Serengeti grassland. They worked together and brought down a large insect by catching the insect by the neck. It was like lions working in group to bring down a large wildebeest, zebra or buffalo.

Again, I noticed that in many cases, the increase in the population of the oribatid mites was followed by an increase in the grass yield. That is in a plot where the population of the oribatid mites was low, the subsequent grass yield was low; where the mites' population was higher, the grass yield was higher, and where the mites' population was the highest, the grass yield was the highest.

In the following year, we carried out the same work in different permanent grassland and we produced the same result.

This finding then raises two questions:

(1) Why did the reduction of the populations of the other groups of the soil organisms lead to an increase in the population of the oribatid mites?
(2) Why did the grass yield increase follow an increase in the population of the oribatid mites?

The answer to question (1) is this:

The oribatid mites and the other groups of the soil organisms were in an ecosystem in which their populations were in delicate balance with one another. In such a system, there are diverse interactions that occur among organisms of the same or different kinds. Through the interactions, pressures arise. These pressures are means of regulating the populations of the groups of the organisms. They prevent the populations of the groups of the organisms from rising above their usual levels in the system. The pressures include predation, parasitism, and competition for food and space. If one or more of these pressures is/are reduced in favour of any of the groups of the organisms in the system, the population of the group will start to rise and the more the pressures are reduced, the more the group's population rises.

So, by reducing the number of the other groups of the soil organisms, the pesticides were reducing the pressures that restricted the growth of the oribatid mites' population. Following the reduction of the pressures, the mites' population started to rise higher.

The answer to question (2) is this:

As we later found out (more details of the finding are in the later part of this chapter), independent researchers identified and reported the ecological roles of the oribatid mites in the system. They reported the mites as extremely important organisms in maintaining soil health and fertility and as having considerable capacity as indicators of soil health and environmental quality. The mites were reported to contribute to organic matter decomposition and nutrient cycling. They break down old materials such as dead leaves and in so doing, they release nutrients in the materials and put them back into the soil. This allows living plants including the grass to pull the nutrients back into their roots so they can grow and feed animals.

So, it becomes obvious that the oribatid mites in the soil ecosystem are directly connected to plants in such a way that an increase in the population of the mites favours plants' growth. But the mites are directly connected to the other soil organisms in such a way that a rise in the populations of the other soil organisms prevents overgrowth of the oribatid mites' population.

In other words, the other soil organisms contribute to regulation of the population of the oribatid mites and through this, they regulate plants' growth. If the population of the oribatid mites rises to a level that is not favourable to the ecosystem, the other soil organisms intensify their pressures, including predatory pressure against the mites' population, and so bring the population back to its usual level. On the other hand, if a circumstance arises which results in a fall of the population of the mites below the usual level, the other soil organisms relax their pressures, making it possible for the mites' population to grow back to the usual level for maintenance of the plants and preservation of the ecosystem.

So, the results of the interconnection, interdependence and activities of the organisms that make up the soil ecosystem are provision of nutrients to plants in an appropriate way, and regulation of the populations of the groups of the organisms for continuity of the ecosystem. This is a proof that: **'the plants supply the needs of the animals, for all the soil organisms, bits and pieces assembled together, mimic the ecosystem in order to supply the needs of the plants'.**

With this finding, I was able to develop a simple formula which one can use to tell which field will produce higher grass yield than the other. This will depend on one knowing the population of each of the groups of the soil organisms in each of the fields before the grass harvest. (The simple formula calculates the proportion or percentage of the oribatid mites in the population of the whole soil organisms. The proportion or percentage correlates with grass yield. Further details about this can be found in the Annex).

Having made the identification, I met the head of the project and told him about it. He was impressed.

However, there was not much information at the time about the role of the oribatid mites in the soil. All that I obtained was information from one textbook which indicated that the mites play a part in maintaining soil fertility. From that information, I suspected that the mites could be decomposers (maintainers of soil fertility).

The head of the project said that the correlation of the grass yield with the oribatid mites' percentage in the population of the whole soil organisms could be because the mites were decomposers, or that it could be just a mere coincidence. He suggested that a new work needed to be done to identify if the mites actually carry out significant decomposition that results in an increase in grass yield.

However, when the two-year project was almost completed, I moved on to new studies. After sometime, I visited the head of the project in Leeds University. It was then that he told me that another work we had done had been published in the Journal of Applied Entomology as previously cited in the biography. He also informed me that he received his Professorship and was leaving Leeds University to take up a new post in the University of Birmingham. He then told me that he had visited North America and had met some scientists who had made a finding that was a confirmation of what I had previously found out and reported to him; that the increase in the grass yield was actually due to an increase in the oribatid mites' population. He said that it was true that the mites were responsible for soil fertility maintenance. He also said that he had told the scientists that somebody in Britain had already made the discovery before them.

Following his return from the North America, the head of the project wrote an abstract of the finding and put my name as the sole author. He showed me the abstract. When I saw my name as the only author, I asked him why his name was not included in the abstract. Traditionally, in a published paper or abstract of a work, the name of the head of the work comes first before the name of any other person who took part in the work. When I asked him why he was breaking the tradition, his answer to me was this:

"I will be the last person to put my name forward for something another person has discovered."

After saying this, he asked me, in a puzzled tone, "But Moses, from where did you get this idea?"

When I received the information from the dream which led to the finding, I kept it to myself because I felt the source (dream) from which it came is not recognised in science. I felt if I made it known, it would be ridiculed and dismissed without further enquiry. Therefore, I decided to investigate on my own so that if it was found true I would declare it.

So, when the head of the project asked me the question, something within me was telling me to declare the truth. I therefore replied that I got the idea from him. He looked at me in suspicion that I was joking. I then went on to complete the story. I told him how in my dream I saw him standing like an angel, holding a sheet of white paper which was too powerful to be handled by ordinary being. I told him how he made it possible for what was written on the paper to reach me, and that what was written on the paper was the idea which led to the finding.

Since then, more findings have been accumulating. They confirm that the oribatid mites were the cause of the increase in the grass yield through their maintenance of soil fertility. Some of the findings and their sources are as below:

One published research has this subheading: **Potential of oribatid mites in biodegradation and mineralization for enhancing plant productivity.**

[4]The research states that oribatid mites decompose litter in the soil ecosystem, leading to nutrient cycling.

Another published research has this subheading: **Oricultural farming practice: A novel approach to agricultural productivity.**

[5]The research states that in oriculture (using oribatid mites in agricultural practice), it was found that oribatid mites significantly increased soil fertility, plant growth and productivity. For example, okra plants cultivated

---

[4] https://dergipark.org.tr/download/article-file/775612

[5] https://typeset.io/pdf/oricultural-farming-practice-a-novel-approach-to-telw3savkl.pdf

by using oriculture farming practice were significantly taller, had flowers earlier, and their pods were larger in comparison with the control plants.

In the next chapter, more information will show that the interconnection and interdependence of organisms in the soil ecosystem are not an isolated occurrence, but essential part of the interconnectedness of all life.

Chapter 2

Interconnection & Interdependence of Soil Organisms and other Lifeforms

As previously mentioned worldview indicates, the life of all organisms on the planet is interconnected and interdependent, forming a wider living system in which the existence of every organism contributes to creating an environment which sustains the existence of the other.

This means that the interconnection and interdependence of the organisms which we identified in our research in the soil ecosystem cannot be an isolated phenomenon, but a part of interconnectedness of all life. In other words, if interconnectedness of all life is visualised as an endless chain, the interconnection of the soil organisms is just a section of the chain. Information from other sources proves that this is the case.

In addition to involving in litter decomposition, soil formation and nutrient cycling, the oribatid mites regulate fungi population by feeding on the fungi.

INTERCONNECTEDNESS OF LIFE

The mites share the job of organic matter decomposition with some other organisms. Examples of these organisms are the fungi and some bacteria.

With this information, it is perceptible that the mites take part in the organic matter decomposition, and also regulate the fungi population by feeding on them. Through feeding on the fungi, the mites keep the fungi population in balance with the populations of the other groups of organisms in the soil.

The above shows that the oribatid mites are not only linked with the other soil organisms we sampled, but also are part of a natural system for organic matter decomposition and regulation of the populations of the organisms that are involved in the decomposition.

This is a demonstration of interdependence of lifeforms, including microorganisms; and indication of interconnectedness of all life.

Again, evidence from photosynthesis shows that plants and other lifeforms (including those in the soil and those above the soil) are interrelated.

Through photosynthesis, plants produce food and oxygen which are essential for their life and the life of animals including man. When animals take in oxygen and food and break down the food, energy is released which the animals use. Carbon dioxide is also released as a waste product. But the carbon dioxide is a raw material which the plants need in making the food through photosynthesis. So the plants take and use the carbon dioxide.

In carrying out photosynthesis, green plants absorb light energy from the sun and use the energy to react with carbon dioxide and water to make sugar called glucose. In the reaction, oxygen is produced as a by-product. Below is a simple equation for the reaction:

Carbon dioxide + water (+ light energy)
= glucose + oxygen.

Animals then eat the glucose through eating the plants. They also breathe in the oxygen. In the animals, the oxygen reacts with the glucose. The result of the reaction is a conversion of the glucose back into the energy, water and carbon dioxide. The animals utilise the energy and breathe out the carbon dioxide as waste. But the plants need the carbon dioxide for making food, and so they take and use the carbon dioxide to make the food (glucose) again. So, on and on goes the cycle of this; give and take – interdependence of animals and plants.

Below is an equation for the reaction of the oxygen and glucose:

Oxygen + glucose = energy (fuel for life) + water + carbon dioxide.

So plants and animals are interconnected in such a way that the life of one depends on the life of the other. The plants produce food and oxygen to keep themselves and animals alive. In return the animals utilise the food and oxygen and release carbon dioxide which the plants need in making the food and oxygen.

Part of the food the plants produce is made available as organic matter to decomposers, such as the oribatid mites in the soil ecosystem. The organic matter includes plant litter and dead animals. Through feeding on the organic matter, the decomposers release mineral nutrients in the matter, making the nutrients available to plants so that the plants can live and produce the food. Thus, the decomposers depend on the plants for their food, and the plants depend on the decomposers for their mineral nutrients.

Bees and other pollinators depend on plants for their food, and plants depend on the pollinators for their reproduction.

Some animals, including some birds, benefit from plants by feeding on the plants' fruits. Through this, they help in dispersing the plants' seeds. The dispersal helps plants' survival because it reduces competition between plants for light, space, water and nutrients.

Evidence shows intimate interdependence of land organisms and aquatic lifeforms.[6]

---

[6] Zimmer, Carl "The Vital Chain: Connecting the ecosystems of land and sea" *Yale environment* 2012 http://e360.yale.edu/feature/the_vital_chain_connecting_the_ecosystems_of_land_and_sea/2529/ (accessed 28/8/15)

Seabirds catch fish as food from the sea and make nests on trees on the land. According to research, the seabirds' droppings (guano) are good fertilizer. This is because of the fish which the seabirds eat. The droppings contribute to mineral nutrients which the plants need for their growth. When the rain comes, it washes some of the nutrients into the sea.

In the sea are microscopic plants known as phytoplankton. When present in high enough numbers, some of them can be seen as coloured patches on the water surface. Just like other plants, they make food through photosynthesis. They also need mineral nutrients for their growth.

By washing some of the mineral nutrients to the sea, the rain makes the nutrients available to the phytoplankton. Just as plants are food to animals on the land, the phytoplankton are food to microscopic animals, small fish and invertebrates in the sea. In turn, the small fish and invertebrates are food to other sea animals. These fish and the invertebrates are also food to some land animals including man.

As one can see hear, the seabirds seem to act like predators that contribute to maintenance of the balance of the populations of the groups of the sea organisms. By feeding on the small fish that graze on the phytoplankton, the sea birds are able to take away some excess fish that may lead to overgrazing of the phytoplankton (reduction of the population of the phytoplankton below the usual level in the system).

INTERCONNECTEDNESS OF LIFE

By means of eating the fish by the birds, the fish become the sustainer of the birds' life as well as a source of plants' nutrients in the form of the birds' guano which the birds drop to the soil.

Washing some of the plants' nutrients to the sea by the rain is a means of transporting the nutrients to the phytoplankton for their growth and therefore maintenance of their population to the normal level in the sea.

So, even though life on land and life in the sea are separated, they are interconnected in such a way that the well-being of one depends on the other.

Based on the information gathered, it is obvious that all life is interconnected to support one another, and the soil ecosystem is the taproot of the interconnectedness.

Plants provide food, oxygen and shelter for animals. The animals release carbon dioxide as an essential part of the raw materials that are needed for the plants to make the food and oxygen for keeping themselves and animals alive. But the organisation and functions of the living things in the soil ecosystem ensure provision of essential nutrients for maintaining the plants' life and for promoting the plants' provision for the animals.

So, without the organisation and functions of the living things in the soil ecosystem, plants and animals could not survive and therefore interconnectedness of life could not exist.

Although every organism plays important role in the interconnectedness of life, this particular role which the organisms in the soil ecosystem play marks the soil ecosystem as the taproot of the interconnectedness of life.

Evidence so far gathered agrees with the worldview that no life exits in isolation, independent of other life. Through their unique characteristics and interconnection, lifeforms are able to contribute to and benefit from the existence of one another.

In the next chapter, it will be visible that the success and well-being of mankind can only be fully achieved by modelling human society on the interconnectedness of life.

Chapter 3

Interconnectedness of Life, a Key to a Better Life and Lasting Success

Some time after our research in the grassland, I was in the process of writing a book about the research outcome. It was during that period that I met a friend, Andy, a postgraduate student at the University of Leeds. I told him about the book and other creative work in which I was engaged, and how these were inspired by the outcome of the research.

Shortly afterwards, I received a letter from the friend. Below is what he wrote in the letter.

"I was reading a book: Collapsing Space & Time: geographical aspects of communication and information, 1991 (eds. Brunn S, Leinbach T), Pg 46. It has a paragraph supporting your model."

He then went on and quoted what is in the paragraph:

"Monocausal models of biological evolution have giving way to conceptual frameworks that recognise that species often coevolve in ways which make it impossible to understand changes in one species without considering simultaneous change in the other species. We will think and teach more effectively if telecommunications and society are viewed as a coevolving complex of processes: telecommunications, societies, economies, governments, human aspirations and human values shape each other, in much the way complexes of living organisms evolved in interaction with each other, each being cause as well as effect of the other's change."

This is evidence that shows that scholars are among the various groups who hold the view that interconnectedness of life is like a textbook of great wisdom, a holder of a key to mankind's lasting well-being and success.

From time immemorial, mystics and some religious groups perceive the importance of interconnectedness of life in man's physical and spiritual well-being, and in the peace of the world.

Buddhists perceive the existence of the interconnectedness of life and its importance. They teach that an understanding of the system can be a key to a more peaceful world and a reduction of sufferings that result from isolation and loneliness. They see the system as a single living whole where all life is interrelated in such a way that the existence of one means the existence of the other; a system where the suffering of one life means the suffering of the other, and the well-being of one means the well-being of the other; a system where man's lack of love has negative effect against him.

INTERCONNECTEDNESS OF LIFE

Man's perception of interconnectedness of life as a created order that reflects the wisdom and glory of God can be deduced from the expression in the Apocrypha – The Bible of Jerusalem: Ecclesiasticus, the glory of God in nature; Chapter 42 verses 21 to 25. Below is the expression:

> "He has imposed an order on the magnificent works of his wisdom,
> he is from everlasting to everlasting,
> nothing can be added to him, nothing taken away,
> he needs no one's advice.
> How desirable are all his works,
> how dazzling to the eye!
> They all live and last forever,
> whatever the circumstances all obey him.
> All things go in pairs, by opposites,
> and he has made nothing defective;
> the one consolidates the excellence of the other,
> who could ever be sated with gazing at his glory?"

This perception is a repeat of the idea that living by the principles of the interconnectedness of life is a route to great wisdom, well-being and glorious success. It also explains the significance of diversity and uniqueness of characteristics of lifeforms. Being different from one another means that every lifeform has something which the other lacks, or something which is opposite to what the other has. Because what one lifeform lacks it can get from others which have it, and what it has it can give to others which lack it, diverse lifeforms bond together forming a system where they live in an interdependent manner; the life of one meaning the life of the other, each consolidating the excellence of the other.

Scholars and mystics agree that if our societies are organised to function like the system of interconnectedness of life, many problems in the world today would not have occurred.

More evidence, as we will see later, shows that a model of the interconnectedness of life can even be a key to achieving spiritual success and eternal life.

Considering the social interaction of members of a cohesive society and the success that follows the interaction, it is apparent that the society is a model of interconnectedness of life. In such society, people of different characteristics interact interdependently, working towards the well-being of one another. The result is improved health, social order and economy.

Being aware of the above, it is now possible to point out the major root cause of the global problems today. In the next chapter, this will be apparent.

Chapter 4

The Tragedy of Failure of Man's Roles in the Interconnectedness of Life and Its Models

Considering the ancient people's beliefs and their attitudes towards nature and wildlife, it seems these people were aware of the interconnectedness of life and its significance. Generally, they perceived animals and plants as sacred entities with which they were intimately connected and upon which their life and well-being depended. For this reason, the people showed great reverence to these lifeforms.

From the distant past, different people perceive the sun as a source of life, a life-giving force, and so they revere it greatly. Some even worship it as a deity.

Today, modern science has proved the authenticity of the perception of the ancient people. It is now clear that man is connected with other animals and the plants. It is the energy from the sun that keeps life on the planet. The sun sends light energy that enables plants to make food for themselves and for other lifeforms including man. Without the sun's light energy, there would be no life on the planet.

Just as other lifeforms have their unique characteristics which are a key to their function in the interconnection, man as part of the system evolved with his own unique characteristics. I believe his characteristics include his intelligence, conscience, instinct to connect and socialise with his fellow man and show love, his creative and artistic skills, his mystical perception and his culture.

These characteristics are for a good purpose. As it is said "From those to whom much is given, much is expected" (Luke 12:48), the characteristics are for man to play special roles in the interconnectedness of life:- for him to function as a manager, a caretaker, an overseer, a supporter, a rehabilitator; for him to conserve; for him to perceive, feel, sense when things are off course and save the situation.

These characteristics shaped man's life and attitude, enabling him to live close to and in harmony with nature. But as he advances in technology and civilisation, some of these qualities continue to atrophy, leading to his steady detachment from nature. The results of this are the increasing pollution, excessive deforestation, overharvesting of plants and overhunting.

Cohesive society, a miniature model of interconnectedness of life, cannot exist without social connection and integration of people in the society. One of the factors that can harm the society is social deprivation. Research shows that social deprivation gives rise to poverty, by which is meant not just financial poverty but also lack of freedom to participate in social activities. This leads to a breakdown of social connection and interaction. The end result is social isolation.

Evidence shows that social isolation leads to problems such as poor personal development; poor social skills; mental illness including depression, anxiety disorder, addiction behaviour; suicide.

Evidence also shows that poor social skills among young people are strongly correlated with their engagement in a variety of risky behaviours, including offending, drug abuse, non-attendance at school, and anti-social attitudes.

One of the reasons people suffer from the above problems is because the natural way of life upon which their well-being depends is taken away from them through social deprivation. Their social connection is stopped, making it impossible for their unique characteristics to work together to support one another for their well-being. This reason is in agreement with the Aristotle's wise sayings below:

> *"Man is by nature a social animal; an individual who is unsocial naturally and not accidentally is either beneath our notice or more than human. Society is something that precedes the individual. Anyone who either cannot lead the common life or is so self-sufficient as not to need to, and therefore does not partake of a society, is either a beast or a god."*

"Whosoever is delighted in solitude is either a wild beast or a god."

"Poverty is the parent of revolution and crime."

The fact that man by nature is a social animal, one whose well-being depends on his connecting and socialising with his fellow man, is well recognised.

Volunteering is one of the ways of social connection and self-confidence improvement. It is a well-known fact that social connection and confidence building improve well-being.[7] As will be seen later, my own work has shown that as the confidence and well-being of participants/volunteers in some social activities improve, and they take up more activities, other areas of their well-being are more likely to improve, leading to a positive cycle of increased personal well-being.

It is through participation/volunteering in activities, for example, activities in a supported (funded) voluntary organisation that one gets connected to people and breaks his social isolation and enhances his well-being. But this channel of well-being improvement can be hampered by poverty.

As said by Aristotle, *"Poverty is the parent of revolution and crime."* Evidence has even shown that poverty is not just the parent of revolution and crime, but also the

---

[7] South, J. "Health and wellbeing: a guide to community-centred approaches." (2015).

parent of deadly illnesses, such as depression and anxiety.[8]

As one may ask, who is then the grandparent of the revolution, crime and deadly illnesses? Obviously, the grandparent is social deprivation. And who is the creator of the grandparent? The creator is not God but man.

It therefore follows that just like God, man has the freedom to create things of his own accord, but unlike God, the result of what he creates in this way reflects his folly – his replacement of love and conscience with selfishness and greed. It shows the extent of his deviation from his natural form and roles – an initiation of a chain of catastrophes against his fellow man and himself.

The saying that 'Man by nature is a social animal' explains a possible reason for evolution of his culture. Culture is one of the unique characteristics of every group of people. It is people's way of life through which they harmoniously connect and socialise with each other and preserve the order of the system of society in which they are.

Culture includes social habits, morals, laws, religions, rituals, food/cuisine, music and art. It is a practice which improves man's well-being, and shapes and regulates his action so that he lives in tune with the principles and order of the interconnectedness of life.

---

[8] Hawton, Annie, et al. "The impact of social isolation on the health status and health-related quality of life of older people." *Quality of Life Research* 20.1 (2011): 57–67.

Research is showing that culture does not only improve the life of people to whom it belongs, but also through cultural diversity, it can enhance other people's culture.

Cultural diversity which is one of the ways of interconnecting people of different cultures and ethnicities is hampered by people's limited insight, prejudice and discrimination. The sad result is a retardation of human progress in life.

Some ethnic groups use musical rhythm as the centre of their culture. Some people perceive this negatively, so negatively that they see it as primitive and a threat.

But recent findings are revealing the importance of musical rhythm, the wisdom behind its place in a culture, and its potential usefulness in facilitating interconnection and well-being of diverse people. One of the findings is as reported in one news release with the headline:

'Symposium looks at therapeutic benefits of musical rhythm.'

To quote the report:

"Musicians and mystics have long recognised the power of rhythmic music. Ritual drumming and rhythmic prayer are found in cultures throughout the world and are used in religious ceremonies to induce trance states. But since the counterculture movement of the 1960s, scientists have shied away from investigating the almost mystical implications of musical rhythm…

"Recent interest in sleep, meditation and hypnosis research has spurred scientists to take a closer look at music. A small but growing body of scientific evidence suggests that music and other rhythmic stimuli can alter mental states in predictable ways and even heal damaged brains."[9]

More reported benefits in this news release and other science journals[10,11,12] which musical rhythm offers include:

- Treatment of a range of neurological conditions including attention deficit hyperactivity disorder (ADHD), and depression, and reduction of chances of developing the problems.
- Reduction of behavioural problems.
- Improvement of intelligence, concentration, learning, IQ, reading and literacy skills, mathematical abilities, emotional intelligence.
- Improvement of cognitive functioning. The musical rhythm was reported to improve cognitive functioning by increasing blood flow throughout the brain. Evidence also suggests that the increase in the blood

---

[9] Saarman, Emily. "Symposium Looks at Therapeutic Benefits of Musical Rhythm." (2006).

[10] Roth, Edward A., and Susan Wisser. "Music therapy: the rhythm of recovery." *The case manager* 15.3 (2004): 52–56.

[11] Roth, Edward A., and NMT MM. "Rhythm for Recovery." Physical Therapy and Rehab Medicine (2006) physical-therapy.advanceweb.com

[12] Surprising Effects of Music – how music affects us and promotes health http://www.emedexpert.com/tips/music.shtml (eMedExper accessed 2/02/2013)

flow can help victims of brain damage regain cognitive function.
- Improvement of recovery from cardiovascular diseases and reduction of chances of developing them.
- Improvement of self-confidence, memory performance, physical performance, body movement and co-ordination.
- Reduction of chronic pain from a range of painful conditions including arthritis.
- Creation and facilitation of positive group behaviour.
- Reduction of stress.

The reduction of stress (increased relaxation) which the musical rhythm offers was reported by other researchers to be effective in keeping some health problems at bay. Evidence shows that chronic stress is among the causes of serious health problems such as heart attack and stroke. It raises blood pressure thereby increasing the risk of heart attack and stroke. It can suppress immune system, contribute to infertility, and speed up ageing process.[13,14,15]

---

[13] Scott, Elizabeth. "Music and Your Body: How Music Affects Us and Why Music Therapy Promotes Health." *Retrieved April* 22 (2007): 2011.

[14] Thaut, Michael, and Gerald McIntosh. "How Music Helps to Heal the Injured Brain." *Dana. org. Np, March* 24 (2010).

[15] Surtees, Paul G., et al. "Life stress, emotional health, and mean telomere length in the European Prospective Investigation into Cancer (EPIC)-Norfolk population study." *The Journals of Gerontology Series A: Biological Sciences and Medical Sciences* 66.11 (2011): 1152–1162.

Chronic stress can lead to mental illness. Mental illness is known as an unhealthy condition of mind which may give rise to cognitive, emotional and behavioural disturbance and physical symptom.

Evidence from my experience that shows the importance of music in culture for social connection and well-being

Musical rhythm is intrinsic in the culture of the Igbo (Ibo) tribe, a tribe in Nigeria where I was born. My mother was a great music composer, story-teller, singer, and music teacher who used music in counseling, healing and comforting people in need. This earned her the nickname: *'Ozigbo Egwu'*, meaning *'the one who teaches Igbos (Ibos) music'*. Another nickname given to her was *'Engine Isimmiri'*, meaning *'an engine, the power and source of reviving water'*.

In the society, from the ancient time, it is believed that music is of mystical origin and that it is a hereditary spiritual gift.

In the past, I asked my mother what inspired her to compose music. She told me that sometimes the music came from a dream, sometimes from a bird's voice, and sometimes it was sparked off by her emotional feeling about the sufferings of people in need. She said that using music to tackle people's problems is more effective than speaking volumes of words.

From my childhood, I witnessed spiritual healers, diviners including some native doctors singing hypnotic songs in carrying out rituals, healing and exorcism, and in communicating with the spirits of the ancestors.

All these indicate that music in a culture has some mystical connections and is important in social order and people's well-being.

My playing musical rhythms and singing songs came to prominence from around 1969. In 1971, I was initiated into a masquerade society. The society is an organisation where masquerades personify the spirits of the dead and of the ancestors who emerge from the earth to join mankind to celebrate special occasions. One of the occasions is the New Yam Festival which is a celebration of yam crop harvest. The period is also a time for the priests of the deities of the land to sacrifice some of the harvested crop to the deities. The sacrifice is to thank the deities for their control of soil fertility, rain, and sun that enable the yam yield, and to entreat them to continue protecting all lifeforms.

The centre of the celebration is music which involves musical rhythm to which the masquerades dance. As multiple musical instruments maintain steady powerful beats whose sound blends with a thrilling sound of high pitched flute known as *oja*, a skillful artist beats a hollowed-out giant wooden drum known as *ikoro*, creating intricate shamanic rhythm as a form of language through which he engages in a dialogue with the energised dancing masquerades. Through the talk and reply, the masquerades release hidden messages and wise sayings that are important in maintaining and improving the order of the society and well-being of individuals.

I was one of those special artists that communicated with the masquerades through playing the ikoro complex rhythm.

I later took a lead in drumming in our church, Ezingodo Methodist Church, to uplift people's spirit in worship. The musical activity in the church developed further, enabling us to be involved in a high profile event to encourage the unity and peace of the nation, Nigeria. The event took place in Enugu, the capital of the then East Central State. I was the chief percussion player in the music of the event.

One of the VIPs in the event who acknowledged the positive effect of our music on the occasion was Sir Francis Akanu Ibiam, a distinguished medical missionary who was appointed Governor of the Eastern Region, Nigeria, from December 1960 until January 1966 during the Nigerian first republic.

In the UK, I led research in a collateral project which involved musical rhythm and we achieved outcomes which are similar to the ones the other researchers reported as shown above.

In Bradford, West Yorkshire, England, I took the lead in a research which involved identifying the effect of musical rhythm on the health and social life of people in need. We carried out this research in schools, Pupil Referral Units, and in communities.

Among the people who participated and benefited in the research were Special Educational Needs (SEN) pupils, young people who were excluded from mainstream school, and older people with some degree of physical/social infirmity and isolation.

The outcomes of the research were identified through evaluations and feedback from the beneficiaries and their carers. Some of the feedback are as follows:

"Then in numeracy, after a wet playtime – usually a difficult lesson with lots of disturbances, e.g. the through traffic in the hall, the SEN children concentrated harder than usual, and got through more work (4 pieces of work) than on a 'normal' day.

"The drumming 'cut out' all other peripheral noises, and the rhythm seemed to help the children to get into the swing of the lesson, more quickly and successfully than normal. Thank you!

"I could now do with a recording of the drumming to use on a regular basis! – Sandra Stuart (SENCo), Crossley Hall Primary School, Bradford."

"The children at Windhill have a very limited experience and understanding of the wider world and its cultures; sadly, their lack of understanding is sometimes expressed inappropriately. I was slightly concerned about the response Moses might provoke when he dressed in costume and began to play, but I need not have worried as all the children were enthralled and wanted to join in.

"The level of engagement and stimulation was total, and the follow-up artwork which many children produced was evidence of the impact the day had on their understanding and creativity.

"For children of similar backgrounds to those of our children, I believe such experiences are essential and I would certainly want Moses to come back to Windhill at a future date. – Mrs. G. Burgess, Headteacher, Windhill Church of England Primary School, Bradford."

"Aireview Centre is part of the Bradford City Pupil Referral Unit based in Shipley. The young people who attend there are excluded from school for various reasons. The main aim of the centre is to re-integrate them back into a new school to continue their education.

"I was asked to provide afternoon activities for these young people. The activities included soccer skills, climbing, art, fitness and African drumming. These activities were based around developing the young people's communication, interaction, fitness and creative skills and also anger management.

"One of the reasons I chose African drumming was because I feel it can bring out these creative aspects through music, especially the drumming. The sessions have proved successful with the young people over the past six weeks especially with the expert and very patient Moses, we have now got to the stage where we are starting to record their music.

"Moses has worked very hard with these sometimes very difficult and challenging young people and I now feel it is producing good results. I am very

glad he agreed to get involved with this project; the young people get on very well with him and have built up a good rapport – Chris Howlett, Education Bradford."

Shortly after this feedback, some of the young people returned to mainstream schools to continue their education and some re-integrated into the community where they now work. Some time ago, I met some of them and they told me that they successfully finished their school and had proceeded to higher education.

Regularly, we do meet some of the young people who now work in the community. In some of the community centres where they work, they applied the skills they gained from the experience to assist us in training other young people in the centres.

The feedback from most of the young people who benefitted from the research is that the drumming workshop changed their life. Some of them said that before, they thought that fighting was a way to solve a problem, but after their development through the drumming workshop that they started thinking and behaving properly.

Evaluation strongly showed that the musical rhythm significantly reduced the depression and anxiety of the older people who were involved in the project. It also showed improvements of other areas of their well-being. The improvements included: reduction of their stress; a boost of their concentration, memory, cognition, co-ordination and body movement; a lessening of their arthritic problems; enhancement of their self-confidence, social skills, and interpersonal relationships; a notable reduction of their social isolation.

Some individuals who were not in need, but who took part in the project as volunteers benefitted as well. An example of their feedback about the benefit they gained is as below:

"After my African drumming session, I went straight to football. I felt so relaxed and it helped me play better. For the first time, we kept a clean sheet and we won!

"My team thinks I should buy an African drum so I can play before every game. – Matthew Reagan, Goalkeeper."

In one of the localities we were carrying out the research, most of the beneficiaries were English people. Following their increasing engagement in the workshop, they developed enough confidence and skill in playing different rhythms. One day, they desired to play rhythm and sing at the same time. They asked me to facilitate this.

Just like many other people, I could not drum and sing at the same time. So I had to develop the technique. I took their song so that I would practise it with musical rhythm. After developing the technique, I taught them what I developed and they learnt it and we rehearsed it together.

Having developed the skill, I started composing simple songs and poems that go with musical rhythms. I then taught the group the songs and poems and they learnt them and we performed these with rhythms.

Evaluation showed that they enjoyed the rhythms and songs when we played the rhythms or sang the songs separately, but they enjoyed it more when we put them together and performed. More enjoyment of performing the rhythms and songs together was common among all the participants, including the elderly, one of whom was 98 years old.

As their confidence increased, one day, they asked me to bring to them the ethnic minority groups in the community to join them so that they would work and perform together. In response, I met the ethnic minority groups. They are groups with different cultures and faiths. They include people of African, African Caribbean, and Asian descent. When I told them that they were needed for social integration with the group I was working with in the research, they were delighted.

"That is what we have been waiting for, but when?" This was their response in general.

So, the building blocks (diverse people with their unique characteristics) for modelling our community on the interconnectedness of life have been established. We are at the moment in the process of building the model.

As one can see here, culture is one of those attributes that man evolved with. Through cultural practice, man is guided to live in harmony with nature and relate with his fellow man. The result is an improvement of his well-being. This was demonstrated in the research which I led. The musical rhythm is a cultural practice which proved itself as a key to social connection; a breaker of social barriers between different groups; an antidote against social isolation and the diseases it breeds. It has proven to be an enhancer of social and life skills that are essential for development of individuals and societies.

Just as every individual is different from another, so is culture. When these differences come together, they perfect each other. The proof of this are the outcomes of the integration of musical rhythm from African culture, and activities such as the unique way of training and developing young people; reducing people's social isolation and improving their well-being in British culture.

It is through this integration that I myself developed further, and became aware of the hidden importance of my musical rhythm skills and the best way to apply the skills to improve my life and the life of other people. At the same time, it is through the integration that the groups with whom I worked became aware of the usefulness of other culture in their own culture and social life. All these were possible because in the integration, we worked in consonance with the principles of interconnectedness of life – we accepted diversity and worked interdependently.

In the next chapter, we will see some evidence that show that religion, a unique culture, evolved as a model of interconnectedness of life for achievement of spiritual needs.

Chapter 5

Lessons Learnt and Insight Gained

Knowledge of the interconnectedness of life reveals the importance of conservation of all lifeforms – protection and sustenance of the life upon which our existence depends. It unveils the significance of diversity and uniqueness of characteristics of lifeforms. It leads to awareness of the roles needed in shaping and preserving a society as cohesive, strong, and healthy.

The knowledge reveals the importance of culture in man's life. It gives an insight that peoples evolved with their cultures as means through which they would live in tune with the principles of the interconnectedness of life, and through this enhance and preserve their well-being.

Evidence indicates that religion, a form of culture, evolved as a model of interconnectedness of life, and through this man would meet his spiritual needs and gain eternal life.

There are strong indications that love is an essential code of practice in religion. It facilitates interrelationships that lead to well-being improvement and achievement of the purpose for which religion evolved.

A proof of the above can be seen in the structure and teachings of some of today's religions such as Buddhism and Christianity.

Buddhists teach that man is related not just to those physically close to him, but also to every other living thing, and through interacting with and relating to others, his existence becomes meaningful. They see one's action as not just affecting other lifeforms including his fellow man, but also affecting himself. For this reason, they teach practice of love. It is through the practice that man's behaviour and actions are shaped and regulated, and his interrelationship with his fellow man is facilitated, resulting in preservation of his physical and spiritual well-being. Their emphasis on the importance of love in mankind's well-being can be deduced from the words of wisdom by Daisaku Ikeda:[16]

> *"When we turn a blind eye to the suffering of others, we grow numb to something important in ourselves and succumb eventually to spiritual paralysis."*

According to the teaching of St Paul, one of the most important figures in the apostolic age, a Christian community is like a model of interconnectedness of life. It is a system made up of individuals of different

---

[16] https://en.wikipedia.org/wiki/Daisaku_Ikeda accessed November 2014.

characteristics, so that with their unique characteristics, each individual will be able to contribute to creating an environment for the well-being of the other, and for achievement of their spiritual goal. The characteristics of the individuals in this case are variety of spiritual gifts. They include the gift of wisdom, gift of knowledge, gift of faith, gift of healing, gift of miraculous power, gift of prophecy, gift of distinguishing between spirits, gift of speaking in different kinds of tongues, and the gift of interpretation of tongues.

Paul said that the particular way in which the gifts are distributed is for a good reason. He likened the individuals with the different gifts to different parts that form one unitary body where they function in an interdependent manner – an interrelationship where a function which one part is not able to carry out, the other parts are; the success of one part becomes the success of the others; if one part is hurt, all parts are hurt with it. If one part is given special honour, all parts enjoy it.[17]

Paul stressed that mutual expression of love among the individuals is a facilitator of the functions of the gifts. To quote him:

"If I have the gift of prophecy, understanding all the mysteries there are, and knowing everything, and if I have faith in all its fullness, to move mountains, but without love, then I am nothing at all..."[18]

---

[17] 1 Cor 12: 4–26

[18] 1 Cor 13:2

He went on to define the characteristics of love:

"Love is patient, love is kind. It does not envy, it does not boast, it is not proud. It does not dishonour others, it is not self-seeking, it is not easily angered, it keeps no record of wrongs. Love does not delight in evil but rejoices with the truth. It always protects, always trusts, always hopes, always perseveres."[19]

The teaching of Jesus Christ, the central figure in Christianity, shows that for one to achieve eternal life, his life needs to be in tune with the principles of interconnectedness of life, and not necessarily with religious observance. This is illustrated in his answer to a question by a man of high status. The man asked Jesus, "What must I do to inherit eternal life?" The answer to the question was that he should love his God with all his heart and love his neighbour as he loved himself.

"And who is my neighbour?" The man asked Jesus. Then Jesus gave him the answer by telling him the parable of the Good Samaritan.[20]

The parable was about a man who on his way was attacked by robbers. The robbers stripped him of his clothes, beat him and went away, leaving him half-dead. One after the other, individuals including a priest who was perceived as an expert in knowing God's law of love approached and saw the wounded man.

---

[19] 1 Cor 13:4–8
[20] Luke 10:25–37

They passed by on the other side of the road, disregarding the man. But a Samaritan, a man perceived as one who did not obey religious law, arrived and saw the robbers' victim. He took pity on him. He helped and revived him.

Jesus then asked the man who was making the enquiry to tell him the one who acted as a neighbour to the wounded man. Was it those who saw him and passed by on the other side of the road, or the one who took pity on him and revived him? The man replied to Jesus that it was the one who showed love and compassion to the robbers' victim. Then Jesus told the man to go and start following the Samaritan's example, for that is what religious law is all about.

By showing sympathy and love to the robbers' victim, the Samaritan proved himself as one who deserved inheriting eternal life. He was the one who was perceived as who did not obey religious law, but his attitude and action proved him as one who did what religion requires. He made this achievement by living his life according to the principles of interconnectedness of life.

In the interconnection, we are bonded not just with those physically close to us but with every living being. In such a bond, diversity is essential. It is a bond where existence and success of one mean the existence and success of the other. It is a bond where if one is hurt, others are hurt with him; a bond where if one turns a blind eye to the suffering of others, he grows numb to something important in himself and succumb eventually to spiritual paralysis.

INTERCONNECTEDNESS OF LIFE

It seems the Samaritan had inner qualities that directed him to perceive these and acted the way he did. In his action he did not consider the wounded man's race, religion, place of origin or whether or not he was related or live close to him. He saw and treated the man as his neighbour, as one of the diverse mankind with whom he was connected.

Through his inner qualities and action, he achieved eternal life and immortalised his name, the name which lives in people's heart even to this day, living as a symbol of a charitable and helpful person – the Good Samaritan.

Understanding of the system and principles of interconnectedness of life reveals that the Good Samaritan had divine insight which is one of the characteristics of humanity, and this guided him to behave the way he did. He perceived that we are all mystically bonded together as a unitary body, a whole where our diversity, uniqueness of characteristics and love work synergistically, yielding beneficial results that make the whole greater than the sum of its parts; a system where the well-being and survival of one enhance and sustain the well-being and survival of the other.

But those who lost the divine insight only perceive diversity and interrelationships as a threat, so, they adopt discrimination and prejudice as a way forward. They try to clone other people in their own image.

Through an insight into the interconnectedness of life, it becomes clear that for one to gain wisdom; rich, peaceful, fulfilling and lasting life, he has to know that prejudice and discrimination are obstacles and counterproductive. He has to know that differences in others are not a threat but assets. They are complements to his own difference in the achievement of the goal.

Knowledge of the interconnectedness of life has led to a better understanding of the saying of the ancient Igbos:

> '*Onye njeghije ka onye isi-awo ihe ama*', meaning
> '*A traveller is wiser than a person of grey hair.*'

It is not the distance the traveller covered that earned wisdom. Rather it is the diverse peoples in their uniqueness that are met through the journey. It is through interacting with these groups of people; through exchange of ideas, knowledge and values with them that led to great wisdom.

Personally, the insight has shaped my perception and attitude to life. Before, I found it hard to acknowledge any other religion other than my Christian religion. But through my insight into the interconnectedness of life, my perception has broadened. I can now see that it is possible that originally all religions evolved to achieve a common good purpose. Just as all beings evolved in their diverse forms and uniqueness for good purpose, so did religions.

In their diverse forms, religions evolved as channels through which people's well-being and spiritual life would be improved. Just as other cultures evolved to connect with and perfect one another, religions evolved in their different forms to connect with and reinforce one another.

I am now aware that it is a mistake for one to despise a religion simply because it looks different, and not because it deviates from the good purpose for which religion evolved. Religion is like a human being whose real nature cannot be fully discerned through his outside appearance, but through his expression of his character. I learnt that to explore a religion before acknowledging or dismissing it is a wise step. I can now understand the wisdom behind interfaith dialogue, and intercultural understanding movement.

These findings raise the questions: How did man perceive the existence of the interconnectedness of life even before the scientific discoveries? Could it be that man by nature is not just a social animal, but also a mystical being and these are part of his innate attributes that enable him to perceive, and manage his life and nature?

Interconnectedness of life, a rough sketch of the system of the kingdom of heaven; a system where if everything in it behaves according to its order, there would be lasting well-being, peace and success; a reflector of the immense wisdom, power and glory of God the creator; a dictionary of every earthly phenomenon and the behaviours of all lifeforms; a textbook of great wisdom which is within our reach, readable by everyone so that we can learn, wisely model our society, regulate, direct and enjoy our life.

The understanding of the system has made it clear that to be able to beat our swords into ploughshares and our spears into pruning hooks, and achieve lasting world peace and order, we need to model the world on the structure and principles of interconnectedness of life; otherwise, we are engaging in a perpetual task with limited success.

Annex
(Report of the Research Work and the Results)

This report is about our finding in our research in permanent grassland. The research started in 1987 and completed in 1989. It took place at the Headley Hall farm, under the auspices of Department of Pure and Applied Biology, Leeds University in the UK.

As indicated in the introduction, it is not necessary to read this Section to understand the book, being just a scientific way of repeating the information in some of the book chapters. It carries information for anyone who wants to know about the overview of the procedures in the scientific research; and about the research data, and the reason the grass yield correlates with the percentage of the oribatid mites in the population of the whole soil organisms. Also in this section are some graphs showing the correlation.

To simplify things, I am going to give the overview of the work, not the details and the precautions taken.

Different forms of pesticides were developed for controlling crop pests. The aim of the research was to investigate the effect of the chemicals on the pests and their natural enemies, as well as on some agricultural soil organisms (microfauna).

In the investigation, permanent grassland was chosen and some plots were created in it. One of the plots was marked as a control (a plot without any chemical in it). Each of the rest of the plots was treated with a different pesticide and named according to the pesticide with which it was treated.

When the chemicals started having their effect, the soil in each of the plots including the control plot was sampled, and every group of organisms (microfauna) in each of the soil samples was extracted and the population counted.

Also, a sample of grass from each of the plots was taken and the pests (certain insects in their life stages) in each of the grass samples were extracted and the number counted. There were some parasitic wasps that control the population of the pests by laying their eggs in the pests. So, we had to dissect the pests from each of the grass samples, extract the parasitic wasps in their life stages which were in the pests and counted them as well. In this way, we were able to determine the numbers (populations) of the pests and the parasitic wasps in the control plot and in each of the plots with pesticide.

After about a month, the grass in each of the plots was harvested and the dry weight of each recorded.

The amount of pest from the control plot was compared to that from each of the plots that was treated with pesticide. The result showed that there was no significant difference between the pest number in each of the treated plots compared to the control plot. In contrast, we found that in many cases, the grass yield from each of the chemically treated plots was consistently higher than that from the control. This was a rather intriguing observation as the number of the pests in each of the plots with the pesticide was almost the same as the one in the control plot. In other words, one cannot say that the lower grass yield in the control plot was because of higher number of pest which reduced the yield.

To solve the mystery, we were thinking of setting up a new project to investigate whether the pesticides were doing something to the biochemistry of the grass. It was during this period that I got a message in a dream which was said to be a route to the solution of the mystery. The message is as below:

> **"The animals supply the needs of man;**
> **the plants supply the needs of the animals;**
> **for all the soil organisms, bits and pieces assembled together,**
> **mimic the ecosystem in order to supply the needs of the plants."**

In the research, I was chiefly involved in sampling the soil in the plots, extracting and counting the soil organisms (microfauna), and recording the populations of the groups of the soil organisms. In response to the message, I looked more closely at the recorded data. We were aware that the pesticides had significant lethal effect on the groups of the soil organisms except one group known as oribatid mites. Generally the pesticides remarkably reduced the populations of the other groups and had little or no effect on the population of the oribatid mites group. Intuitively, I started becoming aware that the oribatid mites' population was infact rising in response to a decrease in the populations of the other groups of the soil organisms. The more the other groups' populations decreased, the more the oribatid mites' population increased. I also noticed that almost every plot where the oribatid mites' population increased had a corresponding increase in grass yield (See table 1 as an example).

Table 1. Data from the control and the propiconazole treated plots.

Plots and the corresponding treatments	Grass yield in tonnes per hectare	Average population of each of the groups of the soil organisms per soil sample		
		Oribatid mites	Mesostigmatic mites	Collembola and minority groups
Control	2.587	13.1	16	57.1
Propiconazole.	2.78	25.4	9.2	36.7

Referring to table 1, in the control (plot without any chemical in it), the average population of the oribatid mites per soil sample is 13.1. The average population of mesostigmatic mites is 16 and the average population of collembola and minority groups is 57.1. The grass yield for this control plot is 2.587 tonnes per hectare.

In the plot with the chemical, propiconazole, the average mesostigmatic mites' population per soil sample is reduced to 9.2. The average population of collembola and minority groups is reduced to 36.7. The result is an increase of the average population of the oribatid mites to as high as 25.4. Following the changes is an increase of the grass yield. In this plot, the grass yield increased to 2.78 tonnes per hectare.

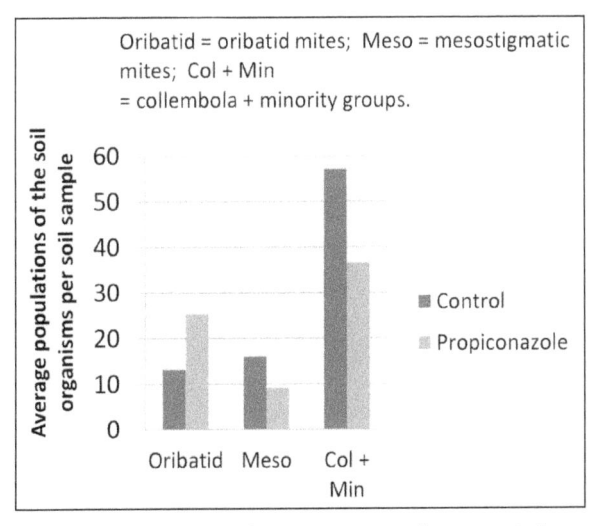

Fig. 1. Bar chart showing the average populations of the groups of the soil organisms per soil sample in the control and in the propiconazole treated plots.

The changes in the populations of the soil organisms can also be seen in Fig. 1. In the control plot, the bar representing the oribatid mites' population is the shortest of all the bars.

In the propiconazole plot, following the decrease of the bars representing the populations of the other groups of the organisms, the bar representing the oribatid mites' population increased remarkably.

The reasons behind these changes are as below:

The oribatid mites and the other groups of the soil organisms were in an ecosystem and their populations were in delicate balance with one another. In such system, there are diverse interactions that occur among organisms of the same or different kinds. Through the interactions, pressures arise. These pressures are means of regulating the populations of the groups of the organisms. They prevent the population of each of the groups of the organisms from rising above its usual level in the system. The pressures include predation, parasitism, and competition for food and space. If one or more of these pressures is/are reduced in favour of any of the groups of the organisms in the ecosystem, the population of the group will start to rise and the more the pressures are reduced, the more the group's population rises.

So, by reducing the populations of the other groups of the soil organisms, the pesticides were reducing the pressures that prevented the population of the oribatid mites from rising above the usual level. Following the reduction of the pressures, the mites' population started to rise higher. In addition, as the mites live under the plant litter, the litter protected them from the chemicals.

As we later found out (please see page 12 for details), other independent researchers identified and reported the ecological roles of the oribatid mites in the soil ecosystem. They reported that the mites are extremely important organisms in maintaining soil health and fertility. In addition, the mites have considerable capacity as indicators of soil health and environmental quality. The mites were also reported to contribute to organic matter decomposition, soil formation, and nutrient cycling. They break down old material such as dead leaves and put the nutrients back into the soil. This allows living plants including the grass to pull the nutrients back into their roots so they can grow and feed animals. This is why the grass yield was high in the plots where the oribatid mites' population was high.

This finding then raises the question: Does it mean that if there are two permanent grasslands, and one has higher population of oribatid mites than the other, then in a month's time, it will also have higher grass yield?

To answer this question, one needs to be aware that in the soil ecosystem, the oribatid mites' population is under the regulation of the other groups of the soil organisms in the ecosystem. If in one grassland, say A, the oribatid mites' population is high and the total population of the other groups of the soil organisms is also high, the oribatid mites' population will face high pressure from the other groups. The high pressure from the other groups will quickly suppress the oribatid mites' population down to the usual level.

In another situation, say grassland B where the oribatid mites' population is equally high but the total population of the other groups of the soil organisms is lower; there will be less pressure on the mites' population, leading to a steady increase of the mites' population even possibly up to the time (a month) the grass would be harvested.

Therefore in grassland B, there would be higher grass yield than in grassland A because in the grassland B, there would be more oribatid mites retained, hence more decomposition of organic matter and provision of nutrients to the grass.

So, to make accurate prediction about which grassland would have higher grass yield than the other, one needs to consider the population of the oribatid mites, as well as the total population of the other groups of the soil organisms which acts as a pressure against the mites' population in the ecosystem.

After putting these factors into consideration, I identified the percentage of the oribatid mites in the population of the whole soil organisms as the appropriate measure. Therefore in a grassland where the percentage is higher, the grass yield would be higher and where it is lower the grass yield would be lower.

For example, in a grassland say A where the population of the oribatid mites per soil sample is as high as say 20, and the total population of the other groups of the soil organisms per soil sample is as high as say 60, the percentage of the oribatid mites in the population of the whole soil organisms will be 20/ (20 + 60) x 100 = 20/80 x 100 = 25.

In another grassland, say B where the population of the oribatid mites is equally 20 per soil sample, and the total population of the other groups of the soil organisms is as low as say 30 per soil sample, the percentage of the oribatid mites in the population of the whole soil organisms will be 20/ (20 + 30) x 100 = 20/50 x 100 = 40.

As the above calculations show, grassland A has lower percentage of oribatid mites than grassland B, so there would be less grass yield in grassland A than in B.

Using this measure or index i.e. the percentage of the oribatid mites in the population of the whole soil organisms in each plot, we found that our predictions were in agreement with our experimental results. (The mites' percentage correlated with the grass yield).

So, the mites' percentage in the population of the whole soil organisms is a valid index to be used in predicting which grassland will have higher grass yield than the other. It represents a measure of the oribatid mites' actual sustainable population increase; and the sustainable population increase is a measure of potential grass yield.

In contrast, identical transformation of our data showed that in the plots, the incremental relationship did not exist between the grass yield and the percentage of any of the other groups in the population of the whole soil organisms. In other words, we calculated each of the other groups' percentage in the population of the whole soil organisms and we found that none of the groups' percentage correlated with the grass yield.

The results from the relevant part of our work are shown in Table 2 and Fig. 2, and in Table 3 and Fig. 3.

Fig. 2 is the graphic representation of the data in Table 2, and Fig. 3 represents the data in Table 3.

(**Key for the tables below:** Oribatid percentage refers to the percentage of oribatid mites in the population of the whole soil organisms; Grass yield refers to grass yield in tonnes per hectare; Population of whole organisms refers to the average population of the whole soil organisms per soil sample; Oribatid refers to oribatid mites; Meso refers to mesostigmatic mites; Col + Min refers to collembola and minority groups).

Table 2. Data from our work in different plots in a permanent grassland in 1987, (The soil organisms sampled on 5/11/1987, and the grass harvested on 2/12/1987).

Plots and the corresponding treatments	Oribatid percentage	Grass yield	Population of whole organisms	Average population of each of the groups of the soil organisms per soil sample		
				Oribatid	Meso	Col + Min
Control	15.2	2.587	86.2	13.1	16	57.1
Methiocab	18.25	2.598	109.6	20	11.2	78.4
MCPA/24DB	21.1	2.66	66.4	14	9.2	43.2
Tricl/Pyrl	23.3	2.69	70.4	16.4	12.4	41.6
Propiconazole	35.62	2.78	71.3	25.4	9.2	36.7

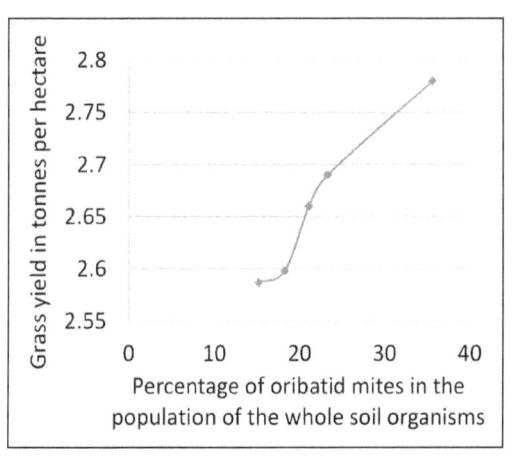

Fig. 2. A plot of the data in Table 2: Grass yield in tonnes per hectare versus oribatid mites' percentage in the population of the whole soil organisms.

Table 3. Data from our work in different plots in a permanent grassland in 1988. (The soil organisms sampled on 25/4/1988, and the grass harvested on 1/6/1988).

Plots and the corresponding treatments.	Oribatid percentage	Grass yield	Population of whole organisms	Average population of each of the groups of the soil organisms per soil sample		
				Oribatid	Meso	Col + Min
Triazophos	15.92	7.388	189.71	30.2	23	136.51
Control	22.15	7.478	218.55	48.4	31.3	138.85
Chlorpyrifos	27.9	7.913	211.45	59	26.2	126.25
Gama HCH	30.83	8.04	158.94	49	22.8	87.14

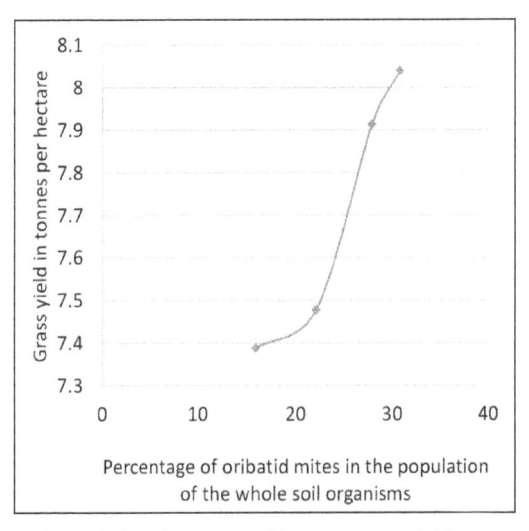

Fig. 3. A plot of the data in Table 3: Grass yield in tonnes per hectare versus oribatid mites' percentage in the population of the whole soil organisms.

In conclusion, our results revealed a corresponding incremental relationship between the percentage of oribatid mites and grass yield of each plot. That is to say that as the percentage or proportion of the oribatid mites in the population of the whole of the soil organisms in each plot increased, the corresponding grass yield also increased. This simple relationship suggests that the activities of the oribatid mites relative to those of the other organisms in the soil ecosystem played a major role in increasing the grass yield. It also correlates with the finding that the oribatid mites decompose organic matter and maintain soil fertility thereby promoting higher grass yield.

Our results also revealed how organisms in the soil ecosystem are interrelated: Oribatid mites promote plants (grass) yield, and the other groups of organisms contribute to regulation of the population of the oribatid mites. The means of the regulation include predation, and competition for food and space.

....

A question one might ask: Does every chemical always spare the oribatid mites and reduce the populations of the rest of the groups of the soil organisms?

Answer: Not always. It depends on the nature of the soil surface. If the soil surface has a shelter that protects any group of the organisms, the chemical will have less or no effect on the group compared to those that are unprotected. Already we know that oribatid mites live under litters including dead leaves. The litters seemed to protect them from chemical. In contrast, some other groups do not have such protection, for example, mesostigmatic mites (the predators) move around in search of their prey and through this easily contact the chemical. So, if any chemical can overcome the barriers that protect the oribatid mites, it would affect the mites.

In a situation where the chemical persistently affects the mites, reducing the population more than it reduced the population of the other groups, the mites' percentage in the population of the whole soil organisms will be relatively lower, resulting in lower grass yield. This is a possible explanation for the change in Table 3. (See the change observed in the plot with the chemical, triazophos when compared to the control plot). Here, the percentage of the oribatid mites in the population of the whole soil organisms is higher in the control than in the plot treated with the chemical, triazophos. The consequence is a lower grass yield in the triazophos plot than in the control. So, the chemical must have acted in a different way which disadvantaged the mites' growth. This includes more lethal effect on the mites.

www.ingramcontent.com/pod-product-compliance
Ingram Content Group UK Ltd.
Pitfield, Milton Keynes, MK11 3LW, UK
UKHW030645300125
454332UK00001B/28